EXPLORING THE MOON

Published in 2007 by The Rosen Publishing Group, Inc.
29 East 21st Street, New York, NY 10010

Copyright © 2007 by The Rosen Publishing Group, Inc.

All rights reserved. No part of this book may be reproduced in any form without permission in writing from the publisher, except by a reviewer.

First Edition

Editors: Melissa Acevedo and Amelie von Zumbusch
Book Design: Ginny Chu

Photo Credits: Cover © Science Source/Photo Researchers, Inc.; p. 4 NASA/JPL–Caltech; p. 6 © Stephen & Donna O'Meara/Photo Researchers, Inc.; p. 8 Calvin J. Hamilton; p. 10 Digital image © 1996 Corbis, original image courtesy of NASA/Corbis; p. 12 © Roger Ressmeyer/Corbis; p. 14 © PhotoDisc; p. 16 © John Sanford/Photo Researchers, Inc.; p. 18 © Mark Garlick/Photo Researchers, Inc.; p. 20 NASA

Library of Congress Cataloging-in-Publication Data

Olien, Rebecca.
Exploring the moon / Rebecca Olien.— 1st ed.
p. cm. — (Objects in the sky)
Includes bibliographical references and index.
ISBN 1-4042-3466-7 (lib. bdg.) — ISBN 1-4042-2174-3 (pbk.)
1. Moon—Juvenile literature. I. Title. II. Series.
QB582.O45 2007
523.3—dc22

2005034248

Manufactured in the United States of America

Contents

The Moon is easy to see from Earth. It is the brightest object in the night sky.

The Moon

Earth is the only **planet** in our **solar system** that has one moon. The planets Venus and Mercury have no moons at all. Mars has two small moons. Its smallest moon is less than 8 miles (13 km) across. Our moon is much larger, at 2,160 miles (3,476 km) across. Four other moons in our solar system are about the same size as Earth's moon. The Moon is closer to Earth than any planet or star is. It is about 239,000 miles (385,000 km) from Earth. It does not make its own light. The Moon is bright because it **reflects** sunlight.

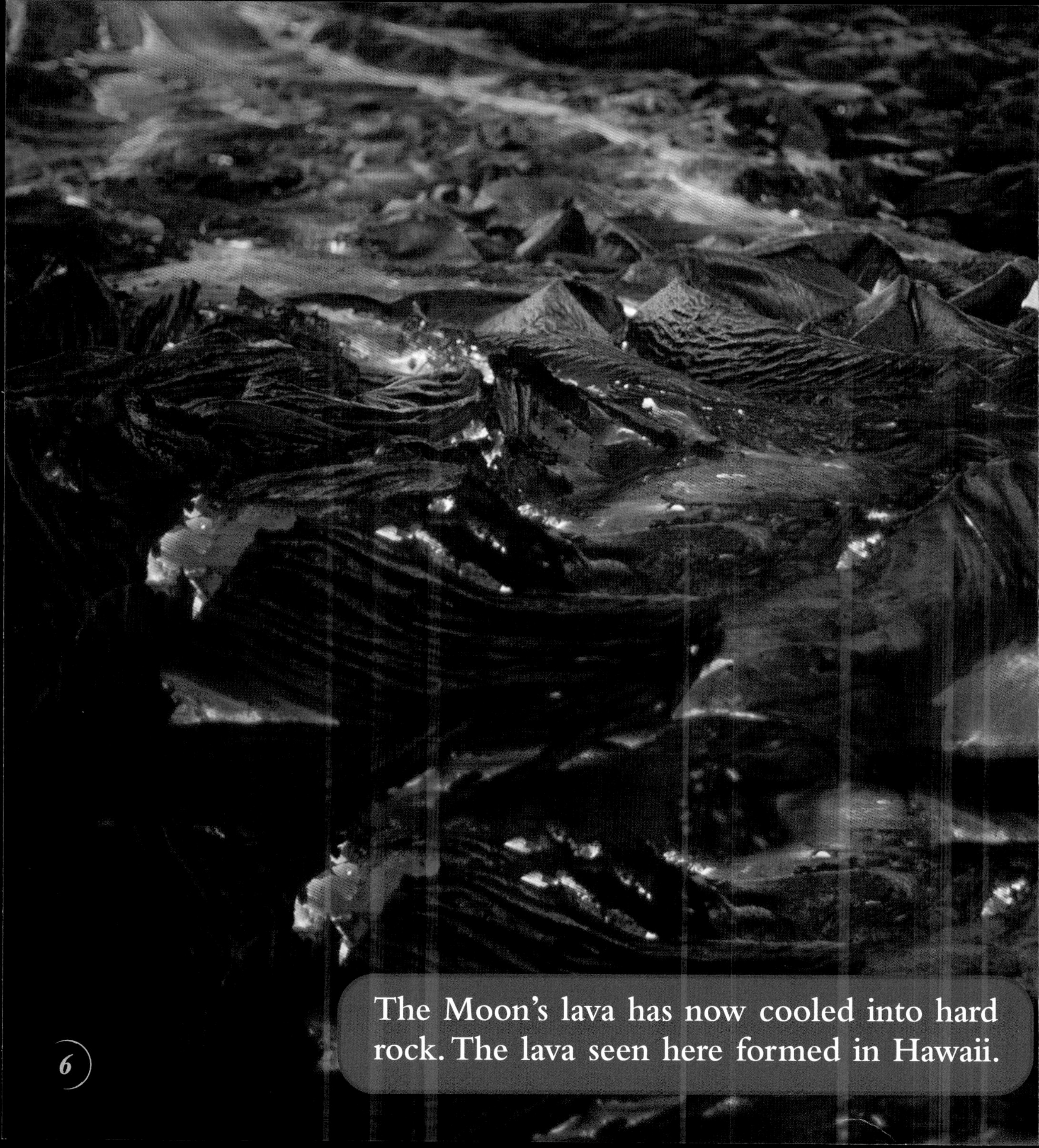

The Moon's lava has now cooled into hard rock. The lava seen here formed in Hawaii.

The Moon's Beginning

The Moon formed about four and a half **billion** years ago. Most **scientists** think the Moon was made when an object the size of Mars crashed into Earth. It struck Earth so hard that huge chunks of rock were thrown into space. Together the object and the rock from Earth formed the Moon. This crash is sometimes called the Big Splat.

The Moon was very hot when it first formed. Its rocks were so hot they melted into **lava**. It took millions of years for the Moon to cool. As it cooled heavier rock settled beneath lighter rock and produced **layers**.

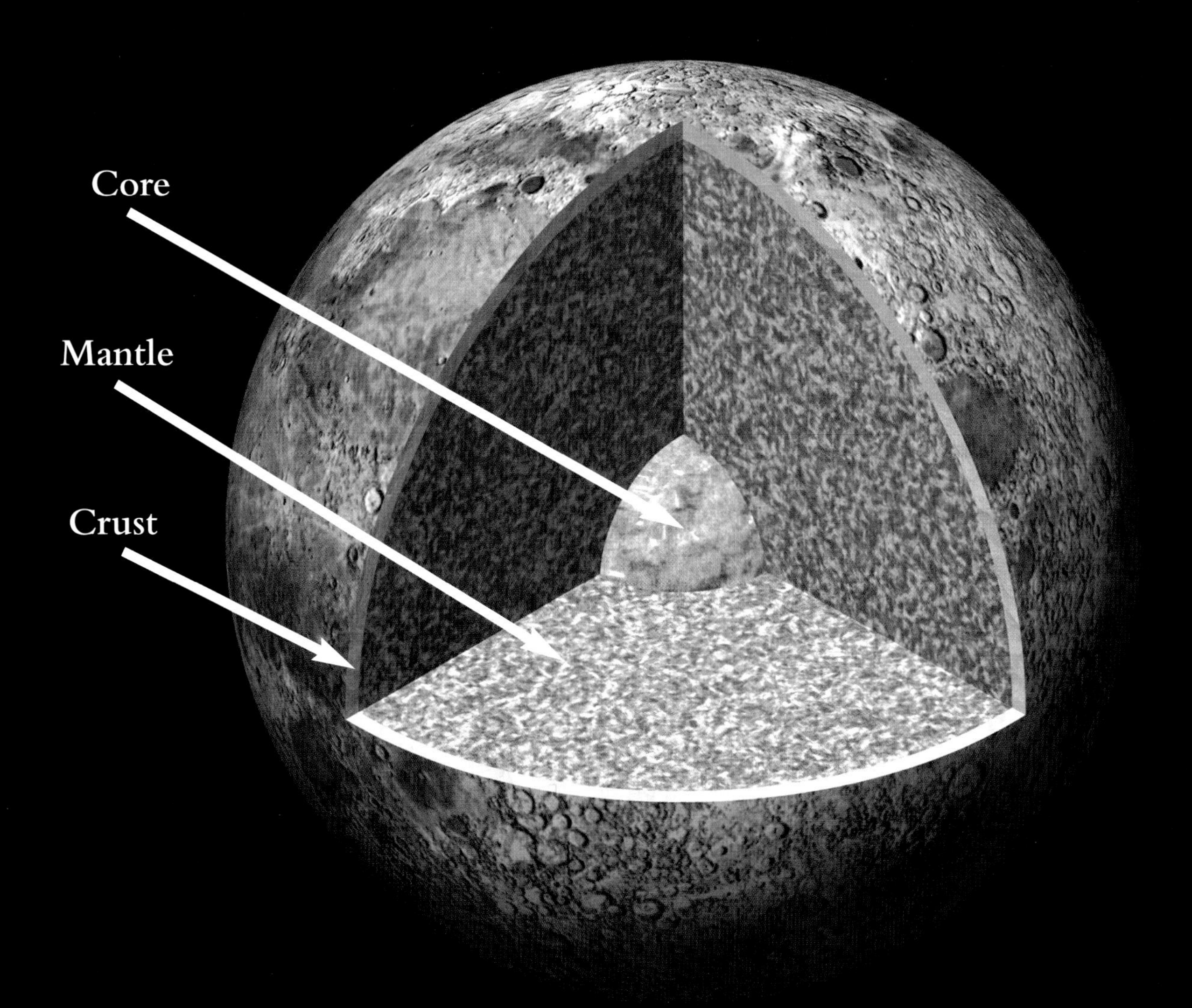

Scientists believe that the Moon's small core has a lot of the metal iron in it.

The Moon's Layers

The Moon is made of three layers, called the crust, the mantle, and the core. The crust is the outside layer. The crust is covered with a layer of dust and small rocks, called regolith. When the Moon formed, the lighter rock floated up to make the crust. The heavier rock formed the mantle.

The mantle is the thickest of the Moon's layers, at about 790 miles (1,270 km) thick. The core is the layer at the center of the Moon. It is the hottest part of the Moon, with a **temperature** of more than 1,832° F (1,000° C).

The Eratosthenes crater, shown here, is 36 miles (58 km) across. It was formed more than one billion years ago.

Craters

A crater is a pit on the Moon's crust made by falling rocks. When the Moon first formed, space was full of moving rocks called asteroids. Many asteroids hit the Moon and dug away land. This left behind craters.

The word "crater" comes from the Greek word *krater,* which means "bowl." The Moon is covered with thousands of craters. Most are the size of small bowls. Others are much larger. The largest crater on the Moon is called the Hertzsprung. It is 369 miles (594 km) across. Many of the Moon's large craters are ringed with tall, rocky walls.

The smooth place here is the Mare Imbrium. It is one of the Moon's maria. Its name means "sea of rains" in Latin.

Mountains and Maria

The Moon has mountains, just as Earth does. They are called highlands. About 80% of the Moon is covered in highlands. The highlands are bright spots on the Moon that we can see from Earth.

Maria are flat places on the Moon. "Maria" is the Latin word for "seas." People used to think the maria were once filled with water. We now know that the maria began as large craters. Lava bubbled up from inside the Moon and spilled into the craters. The lava became hard as the Moon cooled, forming the maria.

The Moon travels around Earth in a path the shape of a somewhat squashed circle.

The Moving Moon

The Moon is a **satellite**. The Moon stays in **orbit** around Earth because of gravity. Gravity is a force that pulls two objects together. It takes a little more than 27 days for the Moon to orbit Earth.

We see only one side of the Moon. This is because the Moon spins as it orbits. It takes the Moon the same time to spin as it does to orbit. The side seen from Earth is called the near side. The side facing away from Earth is called the far side.

This picture shows the Moon's phases. The Sun would be below the bottom of the picture. When the Moon gets bigger it is called waxing. When it grows smaller it is waning.

Phases of the Moon

The Moon goes through eight different **phases**. A new moon happens when the Moon is between the Sun and Earth. During a new moon, the Moon cannot be seen because no sunlight reflects off the side facing Earth. Each night that follows, a little more of the Moon appears.

The Moon looks largest when it is full. It is full when it is on the opposite side of Earth from the Sun. After a full moon, the Moon begins to look smaller each night. It takes about one month for the Moon to go through all eight phases.

During a lunar eclipse, Earth's shadow blocks sunlight from shining on the Moon. In this picture the sunlight is coming from the right.

Lunar Eclipse

A lunar eclipse happens when all or part of the Moon becomes dark. During an eclipse Earth is lined up between the Sun and the Moon. Earth blocks the sunlight, making a shadow that falls on the Moon. In most eclipses Earth's shadow darkens only part of the Moon. This is a partial eclipse.

During a total eclipse, the whole Moon becomes dark and seems to disappear. A total eclipse lasts about 3 hours. During this time people can watch the Moon slowly darken and then brighten as it passes through Earth's shadow.

American astronaut Buzz Aldrin is seen here studying the Moon. Aldrin landed on the Moon on July 20, 1969.

Going to the Moon

The **Soviet Union** sent the first rocket to the Moon in 1959. In 1969, America sent *Apollo 11*. This was the first rocket ever to land on the Moon.

Astronauts took pictures and gathered rocks to bring back to Earth. No sign of life was found on the Moon. The sky on the Moon is always black. This is because the Moon has no **atmosphere** to reflect the sunlight. A total of 70 rockets has been sent to the Moon and 12 astronauts have walked there.

Could We Live on the Moon?

When astronauts visit the Moon, they need to wear special space suits. Even with space suits on astronauts cannot stay there long. This is because the Moon has no atmosphere. Without an atmosphere there is no air to breathe. The Moon also gets much hotter and colder than Earth. Temperatures on the Moon reach 253° F (123° C) during the day. However, the temperatures can drop down to -280° F (-173° C) at night. Some day it might be possible for you to visit the Moon, but it would be very hard to live there.

Glossary

astronauts (AS-troh-nots) People who are trained to travel in outer space.

atmosphere (AT-muh-sfeer) The layer of gas around an object in space. On Earth this layer is air.

billion (BIL-yun) One thousand millions.

lava (LAH-vuh) Melted rock.

layers (LAY-erz) Thicknesses of something.

orbit (OR-bit) The path an object travels around a larger object.

phases (FAYZ-ez) The different shapes of the Moon as seen from Earth.

planet (PLA-net) A large object, such as Earth, that moves around the Sun.

reflects (rih-FLEKTS) Throws back light, heat, or sound.

satellite (SA-tih-lyt) A natural object that circles a planet in space.

scientists (SY-un-tists) People who study the world.

solar system (SOH-ler SIS-tem) A group of planets that circles a star.

Soviet Union (SOH-vee-et YOON-yun) A former country that reached from eastern Europe across Asia to the Pacific Ocean.

temperature (TEM-pur-cher) How hot or cold something is.

Index

Web Sites

Due to the changing nature of Internet links, PowerKids Press has developed an online list of Web sites related to the subject of this book. This site is updated regularly. Please use this link to access the list: www.powerkidslinks.com/oits/moon/